Cells and Cell Function

Carol Ballard

WAYLAND

First published in 2015 by Wayland

Copyright © Wayland 2015

All rights reserved.
Dewey number: 571.6-dc22
ISBN 978 0 7502 9641 0
10 9 8 7 6 5 4 3 2 1

MIX
Paper from
responsible sources
FSC® C104740
www.fsc.org

Produced for Wayland by Calcium

Editors: Sarah Eason and Leon Gray
Editor for Wayland: Julia Adams
Designer: Paul Myerscough
Illustrator: Geoff Ward
Picture researcher: Maria Joannou
Consultant: Michael Scott OBE

Printed in China

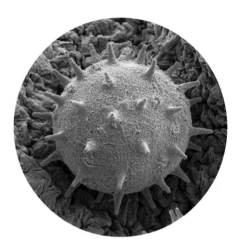

Every attempt has been made to clear copyright. Should there be any inadvertent omission please apply to the publisher for rectification. The author and publisher would like to thank the following for allowing their pictures to be reproduced in this publication:

Cover photograph: Shutterstock/Sebastian Kaulitzki.
Interior photographs: Corbis: Micro Discovery 4, 33, Visuals Unlimited 3, 8, 13, 26, 28, 30, 38; Fotolia: Natalia Sinjushina 37; Getty Images: AFP/Sam Yeh 21, Image Bank/Armando F. Jenik 16; Istockphoto: Jacob Stephens 10; Paul Myerscough: 39; Photoshot: 9; Rex Features: 41; Shutterstock: Adisa 32, Subbotina Anna 27, Chantal de Bruijne 7, Matthew Cole 14, Christian Darkin 12, Philip Date 18, Jiri Flogel 11, Holger W 5, 22, Sebastian Kaulitzki 23, 34, 36, Hugo Maes 17, Mona Makela 25, Mana Photo 35, V. J. Matthew 31, Muellek 40, Alistair Scott 20, S. Game 24, Michael Taylor 6, 45, Veronika Vasilyuk 15.

Wayland is an imprint of Hachette Children's Group
Part of Hodder & Stoughton
Carmelite House
50 Victoria Embankment
London EC4Y 0DZ

An Hachette UK company   www.hachette.co.uk   www.hachettechildrens.co.uk

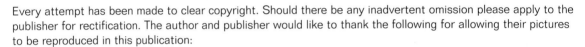

SAFETY NOTE: The activities in this book are intended for children. However, we recommend adult supervision at all times as neither the Publisher nor the author can be held responsible for any injury.

Contents

CHAPTER 1

What is a cell?

Every living thing is made up of cells. Some tiny organisms, called bacteria, consist of just one cell. Very complex animals and plants contain billions upon billions of cells. Cells control the functions of all living things, so they are always very busy and active. Chemical reactions take place inside them all the time.

Some cells live for many years, while others exist for just a few hours or days. New cells are continually being made to replace the old cells as they die. Some cells can move around and change their shape. Some can even engulf and destroy other cells.

Organising cells

Cells are the building blocks for every living organism. When you think about your own body, you probably think about the main parts such as your bones and brain, heart and muscles and eyes and skin. But did you know that they are all made up of cells?

Cells that do similar jobs group together to make tissues. Different tissues are then grouped into larger structures called organs, such as the heart. Organs work together in groups called systems. And all the systems together make a complete organism.

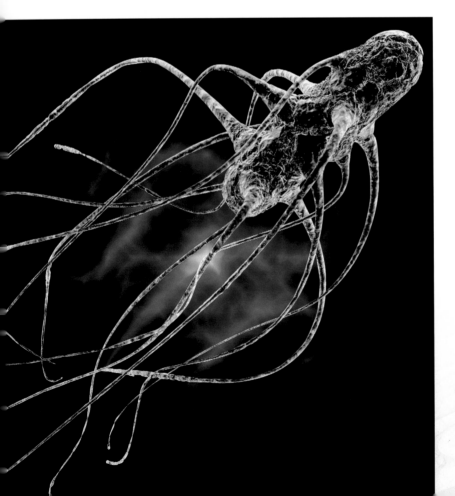

This photograph shows salmonella, a single-celled bacterium that causes food poisoning in humans.

AMAZING CELLS!

Cells come in many different shapes and sizes, and some have amazing properties!

- The cells of some animals, such as fireflies and many deep-sea creatures, carry out chemical processes that make their bodies glow in the dark. This is called bioluminescence.
- The cell of a bacterium called mycoplasma is one of the smallest in the world. If you could arrange the bacteria in a straight line, with each cell just touching the next, it would take more than 30,000 to cover 1mm!
- Some bacteria and single-celled algae have tail-like structures, called flagellae, to help them move around in their liquid surroundings.
- Diatoms are single-celled algae. Many diatoms have beautiful glass-like cell walls made from a hard material called silica. They are often such intricate shapes that it is hard to believe they are living organisms.

Think about the organisation of your own body:
muscle cells ➔ muscle tissue ➔ heart (organ) ➔ circulatory system

What do cells need to function?

All cells need nutrients to provide chemicals and energy. They also must have oxygen to be able to use the energy and carry out important chemical processes.
Water is the third essential ingredient. Water keeps all the structures and liquids inside the cell stable.

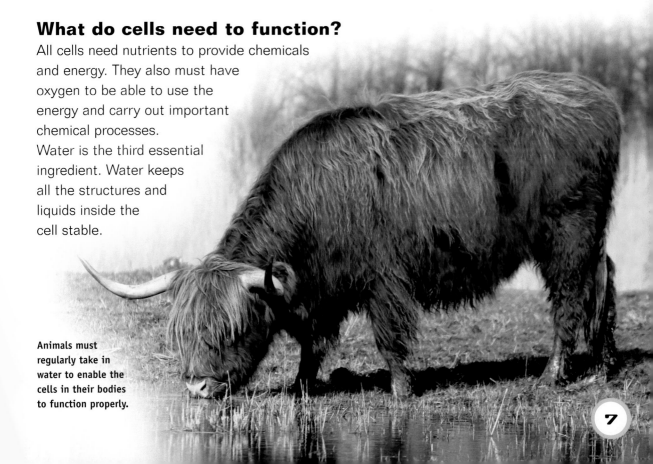

Animals must regularly take in water to enable the cells in their bodies to function properly.

Discovering cells

The first light microscope was developed in the early seventeenth century. For the first time, people could see things that were too small to see with the naked eye. No one knew about cells until 1665, when an English scientist called Robert Hooke (1635–1703) published *Micrographia*. In his book, Hooke described how he had used a microscope to study cork bark. Under the microscope, Hooke saw lots of tiny box-shaped structures in the bark. These reminded him of the small rooms, called cells, in which monks lived. So Hooke called the tiny boxes in the cork bark 'cells'.

Animalcules

Hooke's discovery opened up a whole new world – the incredibly tiny world of the cell. Scientists around the world began to study animal and plant material using their microscopes.

A Dutch scientist called Antoni van Leeuwenhoek (1632–1723) designed and made his own microscopes and examined many animal specimens. In 1676, van Leeuwenhoek observed single-celled organisms, including bacteria and algae. He called them 'animalcules'.

Cell theory

The next big development came from the work of two German scientists, Theodor Schwann (1810–1882) and Matthias Schleiden (1804–1881). Schleiden studied plant cells, and Schwann studied cells in animals. When they compared their work, they came up with an idea that became known as cell theory. This had two main principles:

1. All organisms consist of one or more cells.
2. Cells are the basic structural unit of life.

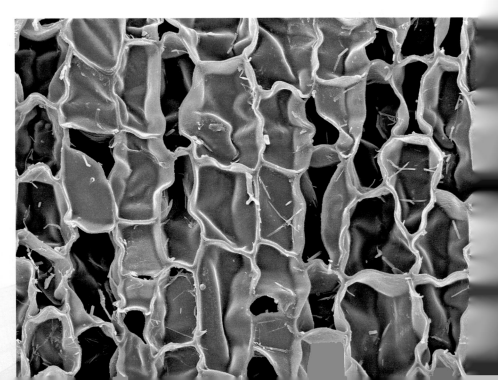

Seen under a microscope, cork cells resemble tiny pocket-like structures.

Theodor Schwann's
pioneering studies
of cell structure and
function are still used
by scientists today.

In 1858, a German scientist Rudolf Virchow (1821–1902) came up with another idea. He said that every cell comes from an existing cell. Together, the ideas of Schwann, Schleiden and Virchow form the basis of modern cell biology.

Look inside

Scientists also began to look inside cells during the nineteenth century. They could see some structures, but the light microscopes they used were not powerful enough to make out a lot of detail. They could not see any detailed structures inside the cells. All that changed in 1931 with the invention of the electron microscope by German scientists Max Knoll (1897–1969) and Ernst Ruska (1906–1988).

TIMELINE

Some important discoveries in the history of cell biology:

1590 The first microscope is invented by Zacharias Jansen and John Lipperhey of the Netherlands.

1655 Robert Hooke observes cells in slices of cork bark.

1676 Anton van Leeuwenhoek observes single-celled organisms that he calls 'animalcules'.

1839 Matthias Schleiden and Theodor Schwann propose the cell theory.

1858 Rudolf Virchow states that new cells can only be formed from existing cells.

1931 Max Knoll and Ernst Ruska develop the first transmission electron microscope.

1935 Knoll describes the idea of a scanning electron microscope.

1938 Manfred von Ardenne builds the first scanning electron microscope.

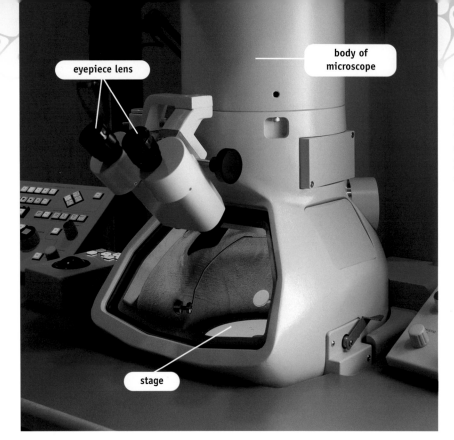

eyepiece lens

body of microscope

stage

Specimens are placed onto the base of an electron microscope, called the stage, and then viewed through the binocular-like part of the microscope.

Viewing objects with a light microscope

Light microscopes magnify objects, making them look bigger than they really are and allowing people to see the tiniest details. There are several different types of light microscopes.

Preparing specimens

A thin slice of material, called the specimen, and a few drops of liquid are put onto a piece of glass called a slide. The specimen must be thin so that light can shine right through it. The specimen is then covered with a 2–3 centimetre square of glass called the cover slip. The slide is then ready for viewing.

Using light

The slide is put on to the microscope stage, which is a flat platform at the bottom of the microscope. Light shines up from a bulb at the base, through a hole in the stage, and through the specimen. The light travels through the objective lens, the main lens in the body of the microscope, through the eyepiece lens at the top of the microscope, and into the observer's eye.

Magnification

Some lenses are stronger than others. Numbers on the metal casing show the strength of the lens. For example, a lens marked x10 will make a specimen look ten times bigger than it really is. To work out the overall magnification, you multiply the strength of the objective lens and eyepiece lens. For example, using a x20 objective lens with a x5 eyepiece lens results in a magnification of 100 (20 x 5). Together, the lenses make the specimen look 100 times bigger than it really is.

INVESTIGATE:
The electron microscope

Electron microscopes can magnify things many times more than light microscopes. Instead of light, electron microscopes use a beam of tiny particles called electrons. There are two main types of electron microscope – scanning and transmission. Look in books and on the Internet for photographs of cells taken using both types of electron microscope. Compare the images. Which type of microscope provides information about the surfaces of specimens? Which provides information about structures inside the specimen?

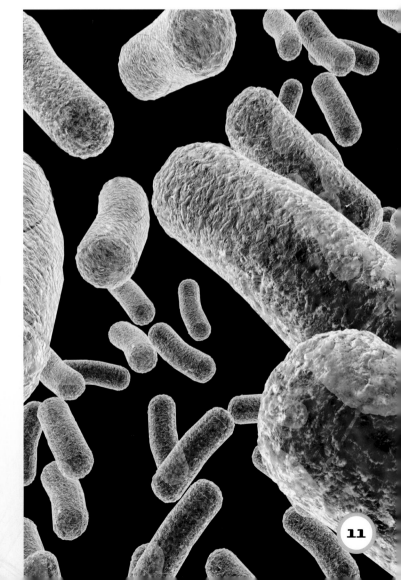

This photograph shows cells observed through an electron microscope. Even fine details, such as the texture on the surface of the cells, can be seen.

Focusing

The focusing wheel moves the stage (and the specimen) towards or away from the objective lens. As the distance between the specimen and the lens changes, you can focus on different depths within the specimen. Most drawings of cells show them as two-dimensional irregular shapes. The drawings just show the view at one depth of focus. If you focus up and down on a specimen that contains cells, you will see that the cells are three-dimensional.

Using stains

Many cells are almost transparent. This makes it difficult to see any detail. A good way to overcome this problem is to add a chemical dye, called a stain, to the specimen. Different dyes attach to different structures in the cell and make them visible.

CHAPTER 2

Inside cells

Cells come in a wide range of shapes and sizes and carry out many different jobs. However, there are some basic structures that are present in virtually every type of cell – from animal and plant cells to single-celled organisms such as algae and bacteria.

Cell membrane

The inside of a cell is contained by a thin outer layer, called the cell membrane. When a cell needs nutrients, the molecules pass through the cell membrane and into the cell. Waste products pass out through the cell membrane in the opposite direction. The cell membrane can also act as a barrier, preventing harmful substances from entering the cell.

Cytoplasm

Inside the cell is a liquid called cytoplasm, which is mainly water with dissolved chemicals in it. Chemical processes take place in the cytoplasm, keeping the cell alive and carrying out other important functions.

Nucleus

Almost every cell contains a nucleus. The nucleus is the site of the organism's genetic material, which is stored in code by a chemical called deoxyribonucleic acid (DNA). The instructions in DNA are called genes. They control everything about the organism, such as all the chemical reactions that take place inside cells.

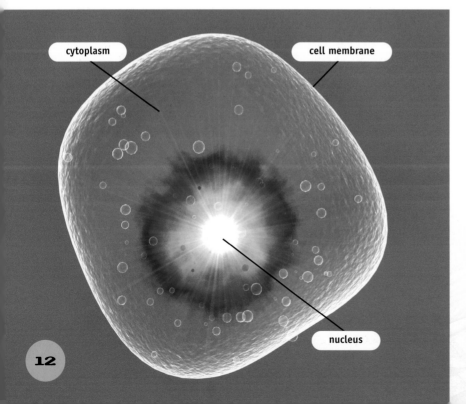

cytoplasm

cell membrane

nucleus

This diagram shows the basic structures of an animal cell.

Genes are also the means by which characteristics are passed from parents to their offspring. Not all cells have a nucleus. An important type of cell that has no nucleus is the human red blood cell. Bacteria and some other micro-organisms also lack nuclei.

Plant cells

Plants have two unique structures in their cells. They are the cell wall and the vacuole.

The cell wall is an extra layer outside the cell membrane. It is made of a strong substance called cellulose, which helps to support the cell and gives the plant strength.

The vacuole is a large space in the middle of the plant cell. It is filled with a liquid called cell sap, which contains water, sugars and salts. The vacuole pushes the cytoplasm against the cell membrane. It is this pressure that helps the plant cell keep its shape.

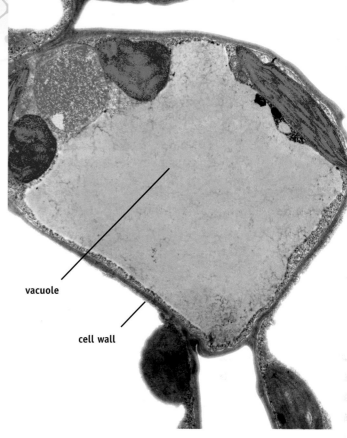

vacuole

cell wall

Plant cells are similar to animal cells, but they also have a cell wall and a vacuole.

CELLS ARE ALIVE!

Under a microscope, the cells in a thin slice of onion skin show up very clearly. The cells appear as neatly arranged boxes, like the bricks in a wall. It looks like nothing is happening, but the cells are very much alive, and many different chemical processes are happening all the time. Streams of chemicals pass in and out of the cell wall. The cells respond to chemicals produced by other cells. Scientists have also detected electromagnetic signals between cells. They think that cells communicate with each other in these ways.

Cell structures

There are many different structures inside cells. Each one has its own name, but together they are known as organelles. Some organelles are found in most types of cells. Others are found only in certain types of organisms or in cells that are designed to carry out specific jobs.

Mitochondria

Most mitochondria are tiny, rod-shaped structures that help break down nutrients inside cells. This process is called cellular respiration. It releases energy for the cell to use. Some cells, such as muscle cells, use up a lot of energy and have thousands of mitochondria. Other cells, such as nerve cells, use less energy and have fewer mitochondria.

Mitochondria are often called the 'powerhouses of cells' because they supply the energy that cells need to function.

Endoplasmic reticulum and ribosomes

The endoplasmic reticulum forms a loose network of folds inside the cell. The tiny ribosomes may stick to the surface of some endoplasmic reticulum. They may also float freely in the cytoplasm. Together, the endoplasmic reticulum and ribosomes make new proteins for the cell which are essential for many cell processes.

Golgi body

The Golgi body is a series of membranes arranged on top of each other like a stack of pancakes. The job of the Golgi body is to build complex chemicals from simple molecules. It stores some of the chemicals for later use and sends some of them out of the cell for use by other cells. The Golgi body also builds organelles called lysosomes.

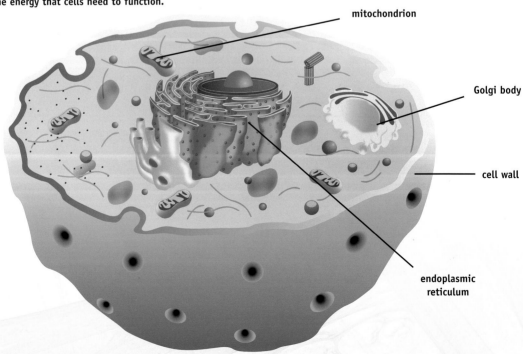

mitochondrion

Golgi body

cell wall

endoplasmic reticulum

Lysosomes

Lysosomes are found in most types of animal cells, but not in plant cells. They are filled with chemicals called enzymes. Lysosomes engulf things such as food molecules, bits of dead cells and harmful chemicals. A lysosome will even engulf another organelle if the cell is short of food! The enzymes inside the lysosome then digest, or break down, the organelle to release energy for the cell.

Chloroplasts

Chloroplasts are found only in plant cells and some single-celled organisms. They contain a green pigment called chlorophyll, which gives the plant its green colour. The chlorophyll traps the energy from sunlight, which the plant uses to make food. This process is called photosynthesis.

Plants rely on chlorophyll, the green pigment found in their leaves and stems. Without it, they would not be able to absorb and utilise energy from the sun.

INVESTIGATE:
Seeing cells

YOU WILL NEED:
onion, kitchen knife, tweezers, microscope and slide, camera

Try looking at some cells for yourself. Cut an onion in half. Carefully peel off one of the thin layers between the stiff, thick layers. Use tweezers to lay it flat on a microscope slide. Add a drop of water and put a cover slip on top.

Lower the cover slip gently to avoid trapping air bubbles. Put the slide on to a microscope stage and look through the eyepiece. Draw what you see. If you have access to a microscope with a camera, you could photograph what you see.

The working cell

Cells are bustling with activity. Hundreds of different chemical reactions are taking place all the time to keep each cell alive. Cells need a constant supply of nutrients, oxygen and water to maintain this activity. They also need to get rid of all the waste products they produce. To do this, the substances have to pass through the cell membrane.

The cell membrane is 'selectively permeable' – it lets some substances pass through more easily than others. Different substances have different ways of crossing the cell membrane.

Diffusion

The molecules in a liquid are moving about all the time. They spread themselves out so that they are evenly distributed throughout the liquid. This spreading-out process is called diffusion. Some smaller molecules, such as carbon dioxide, oxygen and water, can pass through the cell membrane very easily. They move by diffusion until they are evenly distributed inside and outside the cell. The cell has no control over this process.

Osmosis

Living organisms contain solutions in which molecules such as salts and sugars are dissolved in water. If a solution contains more dissolved salts and sugars on one side of the cell membrane than the other, water molecules move through to restore

Sharks absorb seawater by osmosis. Bull sharks, such as the one seen here, can regulate the levels of salt in their bodies so that they can survive in both freshwater and saltwater.

the balance (the sugar and salt molecules are too big to pass through the cell membrane). This process is called osmosis. If too much water leaves a cell, it will shrivel up. This is why some plants are killed in salty seawater. If too much water enters a cell, it may swell and burst.

Active transport

Very large molecules, such as proteins, cannot pass through the cell membrane by diffusion or osmosis. Instead, chemicals called carriers stick to the molecules, help them through the cell membrane, and then release them on the other side. This process is called active transport and uses up energy in the form of a chemical called adenosine triphosphate (ATP; see box).

Mangrove trees survive in saltwater because they take up the saltwater by osmosis, which means that the salt is unable to enter the plant tissues. In this way, the plant obtains fresh, non-salty water.

STORING AND RELEASING ENERGY

Two very similar chemicals store and release the energy in cells. They are adenosine diphosphate (ADP) and adenosine triphosphate (ATP). ADP becomes ATP when an extra phosphate group attaches to the ADP molecule. This process also locks energy into ATP. ATP can go anywhere in the cell where energy is needed. The ATP molecule then releases the extra phosphate group, turning it back into ADP. The locked-in energy is also released.

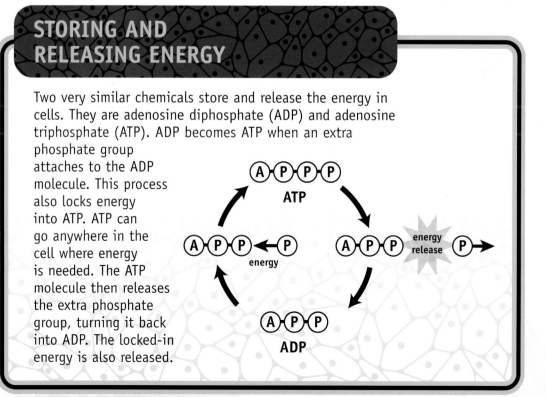

Making new cells

Cells continually wear out and die, so they must be replaced all the time. This is achieved by cells dividing to make new cells. First, the genetic material in the cell is copied. Then the cell splits into new cells, each with the same genetic material as the original cell.

DNA and chromosomes

The nucleus is the site of the organism's genetic material, which is stored in a chemical called deoxyribonucleic acid (DNA). The DNA is contained in thread-like structures called chromosomes. The instructions in DNA are called genes. The different genes control different characteristics. For example, one gene might control whether a person has blue or brown eyes. Another might control whether their hair is curly or straight.

Characteristics such as skin colour and hair colour are determined by DNA.

The complete set of genes is known as the organism's genome. Each organism has a unique genome. The number of chromosomes and the order of genes in the DNA is the same for every individual of the species. All that changes is the characteristic controlled by each gene. For example, all maize plants have 10 pairs of chromosomes. A gene on chromosome 2 controls the kernel colour. Plants with one variant of this gene have yellow kernels. Plants with a different variant produce brown kernels.

Mitosis

The process by which one parent cell splits into two new cells is called mitosis. Each chromosome has two strands that are held together in the middle. As mitosis begins, the membrane around the nucleus disappears.

INVESTIGATE:
DNA model

YOU WILL NEED:
black and yellow beads,
flexible wire, wire cutters

DNA is a complex twisted spiral called double helix. Try making a model of DNA using beads and wires. Use the diagram below right as a guide:

1. Measure the length of two beads, add 2 centimetres and then cut 20 pieces of wire to this length. Cut another 2 pieces of wire 11 times the length.
2. Thread a black bead on to each of the two long wires. Bend the end of each wire to stop the beads falling off.
3. Thread one yellow bead on to one of the short wires and attach one end to a long wire just above the black bead. Do the same with the other end.

4. Add another black bead to the long wires and two yellow beads to three short wires. Attach the short wires to the long wires, placing a black bead on the long wires between each short wire. Now attach a short wire with one yellow bead. Carry on adding beads to the wires in this pattern until you have reached the ends of your long wires. Bend the wires to keep the beads in place. Hold the bottom of the model still and gently twist the top into a spiral.

The two strands of each chromosome are pulled apart and move to opposite ends of the cell. A new membrane forms around each set of chromosome strands. The cell membrane then begins to 'pinch in' at the middle, dividing the parent cell into two new daughter cells – each with a complete set of chromosomes in the nucleus.

Meiosis

Male and female sex cells, such as egg and sperm in people, are produced by a different process, called meiosis. Meiosis is similar to two cell divisions by mitosis, one after the other. Meiosis results in a single parent cell producing four daughter cells – each with a half set of chromosomes. When the male and female sex cells unite during sexual reproduction, the resulting cell (the early stages of a new life) has the full set of chromosomes.

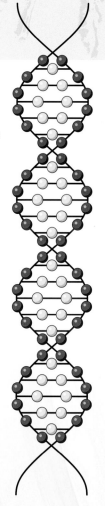

DNA controls the appearance and function of every organism on Earth.

Changing cells

Cell division does not always produce genetically identical daughter cells. Mistakes can occur when the cell makes copies of its chromosomes. These mistakes cause changes, called mutations. There are five main types of mutations:

- **deletion** – a piece of chromosome is missing
- **duplication** – a piece of chromosome is copied twice
- **inversion** – a piece of chromosome is copied back to front
- **insertion** – a piece of one chromosome is copied on to a different chromosome
- **translocation** – two chromosomes swap pieces

Some mutations have no noticeable effect on an organism. Others are helpful to the organism, such as the mutation in many bacteria that help them resist antibiotics. Many mutations are harmful and can lead to cell damage and death, as well as the development of diseases and disabilities. Gene therapy is a recent medical development that offers a way of treating the diseases caused by mutations.

What causes mutations?

Some mutations occur naturally as a result of mistakes in the copying process. Others are caused by exposure to toxic chemicals or radiation. Drug companies carry out thorough tests to ensure that new medicines do not damage chromosomes. Other chemicals that might affect people, such as those used in foods and paints, are also tested.

Some scientific reports suggest that certain chemicals used in industry and agriculture may be linked to genetic mutations.

Scientists used genetic modification to alter the genetic make-up of this mouse so that it glows in the dark.

Genetic modification

Scientists have mapped the genomes of many different organisms, including humans. Genetic modification is a technique that allows them to transfer a gene from one organism to another. This is a sort of artificial mutation. For example, scientists transferred a gene from a fluorescent jellyfish into a zebrafish to create an organism they called a glofish. The glofish was still a zebra fish, but the jellyfish gene made it fluorescent.

Scientists can use genetic modification to produce crops that have advantages over existing crops, such as resistance to drought. However, many people worry about whether or not GM crops are safe.

RADIATION

Radiation such as gamma rays, ultraviolet radiation and X-rays can damage DNA. This can damage the chromosomes of individual cells. Luckily, cells can usually repair any damage. However, if the damage is severe, the cell may die. In some cases, the damage interferes with the process of mitosis. The mutated cells divide uncontrollably, resulting in the illness we call cancer. Exposure to large doses of radiation is very dangerous. For example, the incidence of cancer increased significantly in Japan after atomic bombs exploded in Hiroshima and Nagasaki in 1945, and in Chernobyl, Ukraine, after the accident at a nuclear power station in 1986.

Life from cells

What is the difference between a living creature and something that has never been alive? Apple trees, butterflies, dolphins and polar bears are all living things. They carry out the basic life processes of breathing, sensing, moving, growing, feeding, reproducing and excreting. Although rocks, clouds, metals and icebergs are all found naturally in the world around us, they do not carry out the basic life processes and so they are not alive.

Viruses

All living organisms are made up of one or more cells. Living things carry out basic life processes, and they contain genetic information in the form of DNA. Do viruses have these characteristics? Well, the answer is yes and no.

All living things, such as this butterfly, are made from cells.

Viruses are microscopic particles. They have an outer protein coat that is similar to a cell wall. Viruses contain genetic information in the form of ribonucleic acid (RNA), which is very similar to DNA. However, viruses do not contain cytoplasm or organelles like most cells. Although viruses can reproduce, they can only do so by using a host cell. They do not carry out the other life processes. For these reasons, viruses are usually regarded as non-living particles.

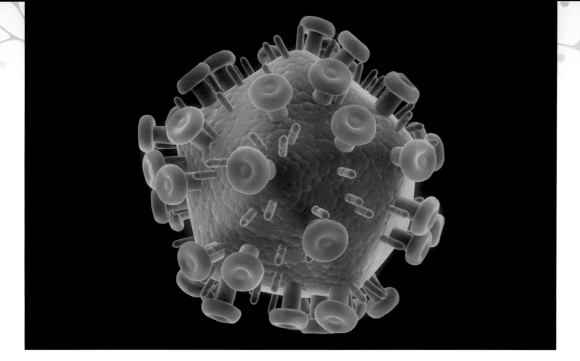

Virus particles, such as this HIV virus, are not living cells.

Single-celled organisms

Single-celled organisms may be very simple, but they can still sense the world around them. They can feed and many can move. Some single-celled organisms can communicate, and many live in colonies.

There are three main types of single-celled organisms: monera, protists and yeasts:

- Monera do not have a nucleus. Instead, their genetic material floats around in the cytoplasm inside the cell. Bacteria and blue-green algae are examples of monera.
- Protists share features of both plant and animal cells, including a nucleus. Amoebas and slime moulds are examples of protists.
- Yeasts are a bit like plant cells, but they do not have chloroplasts. As a result, yeasts cannot make their own food by photosynthesis and get their nutrients from other sources. Scientists classify yeasts as fungi, along with mushrooms and moulds.

HOW DO SINGLE-CELLED ORGANISMS FEED?

Different single-celled organisms feed in different ways. Some are parasites, which means they live on or inside a host and get their food from it. Others feed by engulfing other living organisms and digesting them. Some single-celled organisms make their own food by photosynthesis, converting carbon dioxide and water into food using the energy from sunlight. A few use minerals and water to make food in a process called chemosynthesis.

Monera

Scientists divide the monera into two groups. One consists of the bacteria, the other consists of the cyanobacteria.

Bacteria

Bacteria are probably the most common type of living thing on Earth. There are many different types, and they come in many different shapes and sizes. Most are rod-shaped, but others are spherical, spiral or bent in the shape of a comma. Most bacteria measure less than one thousandth of a millimetre long. Bacteria have cell walls, cell membranes and cytoplasm, but they lack nuclei. Instead, the DNA floats freely in the cytoplasm. Tiny rings of DNA, called plasmids, may also be present. A few bacteria have slime coats and tail-like flagellae to help them move around.

Most bacteria feed on other living things or on plant and animal waste. They produce enzymes to break down the food before absorbing the nutrients.

Bacteria reproduce by a process called binary fission, in which a single cell splits into two identical daughter cells.

DNA (red) can be seen in the cytoplasm in this bacterium.

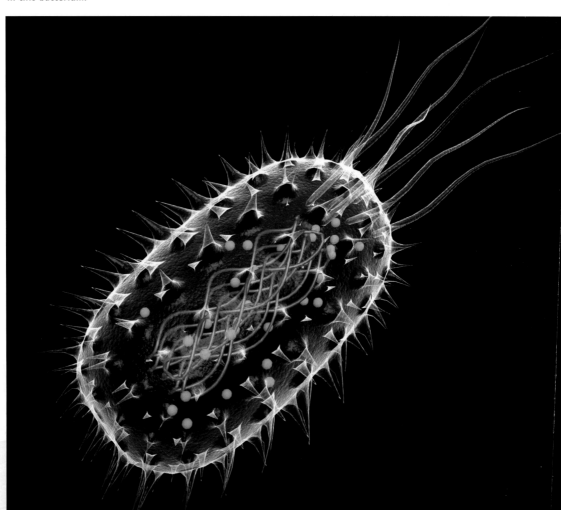

INVESTIGATE:
Bacteria and humans

Bacteria are everywhere. They are found on every surface we touch, and they live on our skin and inside our bodies. Some bacteria are helpful, but many are harmful and cause diseases such as cholera. Anti-bacterial products are available to the public. These kill bacteria and protect us from disease. How many anti-bacterial products are in your house? Check if the products are anti-bacterial by looking at their labels then record your findings in a chart.

Product	Where found	Used for

Cyanobacteria

Cyanobacteria are also called blue-green algae. Until recently, scientists classified them as plants but they are now put in the Kingdom Monera, along with the bacteria. Cyanobacteria have been around for a very long time. Scientists have found fossils dating back more than 3 million years. Most cyanobacteria live in water (and are an important source of food for marine creatures), but some are found in soil on land. They move by gliding over surfaces.

Cyanobacteria contain chlorophyll and another pigment called phycocyanin, which gives them their blue-green colour. Scientists think that chloroplasts in plants evolved from ancient cyanobacteria. Cyanobacteria can also 'fix' nitrogen in the air. They use it to build complex molecules.

Cyanobacteria can be useful to people. Some types naturally produce hydrogen. Scientists are looking to use these organisms as a renewable energy resource.

Spirulina is a cyanobacterium that is often used as a food supplement.

Protists

There are many different types of protists. Most live in water or in the body fluids of other living things. Protists are usually very small, but some are big enough to be seen by the naked eye. Some protists share the features of plant cells. These protists contain chloroplasts and make their own food by photosynthesis. Euglena and Chlamydomonas are plant-like protists. Others share the features of animal cells and find their food from other sources. Amoeba and Paramecium are animal-like protists.

Amoeba

Amoeba is the name of a group of protists that move around as the cytoplasm inside their cells moves in a particular direction. Their shape constantly changes as they move. Amoebae collect their food by engulfing bacteria and other microorganisms.

Paramecium

Paramecium live in water. They are covered in tiny hairs, called cilia, that beat regularly to help the cells swim. A band of cilia wafts water towards a groove in the cell through which the Paramecium feeds.

Euglena cells

Euglena are long, thin cells with a tail-like flagellae that whip through the water to aid movement. Euglena contain chloroplasts, so they can make their own food by photosynthesis. A light-sensitive area, called the eyespot, at one end of

AMOEBA

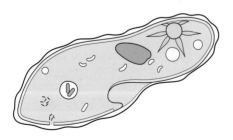

PARAMECIUM

The tails of these Euglena can be clearly seen here. The cells use their tails to propel themselves forward.

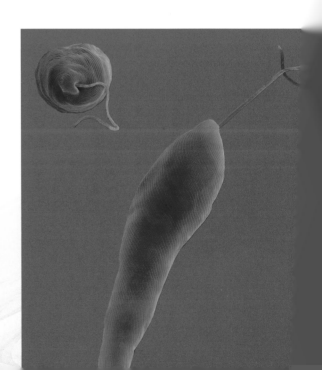

MALARIA

Malaria is a deadly disease that affects people in tropical parts of the world, such as sub-Saharan Africa. The disease is caused by plasmodium protists and spreads through mosquito bites. When a mosquito bites, the plasmodium cells pass into the blood. There, the plasmodium protists reproduce inside red blood cells. The red blood cells eventually burst, spilling out more plasmodium protists to infect more red blood cells. This results in the symptoms of malaria.

Malaria, which is caused by the parasite plasmodium falciparum, is spread from person to person by infected mosquitoes.

the Euglena helps them to find light to photosynthesise efficiently. Euglena also absorb food from the water in which they live.

Chlamydomonas

Chlamydomonas are egg-shaped cells with two long flagellae that help them move in the water. Like Euglena. Chlamydomonas contains chloroplasts to photosynthesise and has a light-sensitive area.

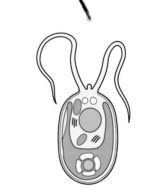

CHLAMYDOMONAS

Diatoms

Diatoms are a group of protists with cell walls made of the mineral silica. Under a microscope, many of these microorganisms have beautiful and ornate shapes. Diatoms live in water and make their own food by photosynthesis.

DIATOM

Yeasts

Scientists have identified more than 1,500 types of yeasts. They live in many different habitats – on plants, in the soil and inside the bodies of animals. Some yeasts are harmful to people and cause a range of infections collectively known as 'thrush'.

Yeast cells are usually egg-shaped or spherical. They have a cell wall and a cell membrane. There is a large vacuole inside the cell, and the cytoplasm contains mitochondria and other organelles. In many ways, yeast cells are similar to the cells of green plants. There is one important difference – yeast cells do not contain chloroplasts and cannot make their own food.

How do yeasts feed?

Yeasts feed on sugars and other complex chemicals. They get these nutrients from sources such as plant saps and dead skin cells. Yeast cells produce enzymes that break down sugars, releasing energy for the yeast cell to use. This process also produces carbon dioxide and alcohol. Both products are used by people. In bread making, yeasts produce carbon dioxide that helps make the bread rise.In brewing, yeasts produce the alcohol for beers and wines.

The tiny blobs on some of these yeast cells are buds. These will develop into new cells.

Living conditions

Yeasts need water, warmth and air to reproduce. They grow best at temperatures of 30°–37°C, but they can survive in warmer or cooler conditions. Most yeasts are killed if the temperature rises above 50°C.

How do yeasts reproduce?

Yeast cells reproduce by a process called 'budding'. A small lump or bud appears on the parent cell. As the bud grows, the nucleus of the parent cell divides by mitosis to form two new nuclei.

One moves into the bud. Eventually, the bud separates from the parent cell to form an identical daughter cell.

Yeast cells also reproduce using spores. Meiosis produces four daughter cells, called spores, each with a half set of chromosomes. When two spores combine, they create a new yeast cell with a full set of chromosomes.

INVESTIGATE:
Living yeasts

YOU WILL NEED:
4 bowls, warm water, sugar, teaspoon, dried yeast

Yeasts are living cells that need water, warmth and energy. What happens when they are deprived of them?

1. Put fresh or dried yeast into four bowls.
2. Add 100ml of warm water to three bowls.
3. Add a teaspoon of sugar to the fourth bowl.
4. Add a teaspoon of sugar to two of the bowls with warm water.
5. Stir the mixture in each bowl with a clean spoon.
6. Put one bowl with water and sugar in a fridge.
7. Put the remaining bowls in a warm place. Leave for 30 minutes.

Results
1. The bowl with warm water and sugar that was left in a warm place should be frothy. The yeast cells had warmth and water, so they released carbon dioxide, a gas which made the water frothy.
2. The bowl with only warm water that was left in a warm place and the bowl with warm water and sugar that was put in the fridge should be slightly frothy or not frothy at all. The yeasts cells did not produce carbon dioxide.
3. The bowl with yeast and sugar that was left in a warm place lacked the water necessary to produce carbon dioxide, so should not be frothy.

Plant cells

Plant cells are enclosed within a rigid cell wall. A cell membrane encloses the contents of the cell. The cell contains a fluid, called cytoplasm, and a large space called the vacuole. The nucleus and other organelles, such as mitochondria and chloroplasts, float around in the cytoplasm.

Plant cells come in many different shapes and sizes. Many are adapted to carry out specific jobs.

Palisade cells

Palisade cells are found only in the leaves and stems. These cells are very efficient at carrying out photosynthesis. They can do this because:

- they are usually in the upper part of a leaf or the outer part of a stem to absorb as much sunlight as possible;
- they contain a large number of chloroplasts, which trap the energy from sunlight;
- within each palisade cell, more chloroplasts are arranged in the outer part of the cell than the inner part to trap as much energy as possible.

Root hair cells

Root hair cells are found just behind the root tips of a plant's roots. They are adapted to absorb water and minerals from the soil. Root hair cells are long and thin and grow outwards from the root. They have a large surface area in contact with the soil. They also have a thin cell wall through which water and nutrients can pass easily into the root. Root hair cells contain no chloroplasts. Light cannot penetrate the soil, so photosynthesis would be impossible.

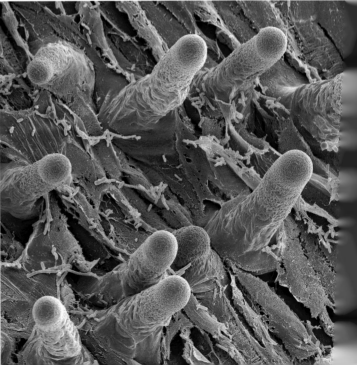

The long blue structures in this photograph are the root hair cells of a plant root.

INVESTIGATE:
Chlorophyll and light

YOU WILL NEED:
kitchen towel, 2 saucers,
bean seeds, saucepan

Discover how photosynthesis works.

1. Slightly moisten two sheets of kitchen towel and place on to two saucers.
2. Put five bean seeds on each towel – with some space between each bean.
3. Put both saucers in a warm, sunny place.
4. Cover one saucer with a light-proof container (such as a saucepan).
5. Moisten each paper towel daily. Check the seeds at the same time, but try to keep the 'darkened' saucer as dark as possible.
6. Compare the saucers after one week.

Results
You should find the beans in the darkened saucer are white and weak. The beans in the saucer that was exposed to the light should be green and strong. Without light the seeds in the darkened saucer could not make chlorophyll, and so could not photosynthesise to create food to grow.

Xylem cells

Xylem cells make up part of a plant's transport system, which is designed to carry water around the plant. The xylem cells line up end to end to form a column, and the walls between the cells break down to form a long tube. Water could not travel very far in a single xylem cell, but it can travel around the whole plant through the interconnecting system of xylem cells.

The green parts of these leaves contain more chlorophyll than the white parts. More photosynthesis can therefore take place in the green parts.

Making seeds

A flowering plant develops from a seed. The seed itself develops when a male sperm cell found in a pollen grain fertilises a female egg cell, or ovule. Both the sperm and egg cell carry half the normal number of chromosomes. When the two cells join during fertilisation, they combine to make the full set of chromosomes in the seed. Fertilisation occurs when a pollen grain is carried from one flower to another by the wind or by an animal such as a bird or an insect.

THE REPRODUCTIVE ORGANS OF A FLOWER

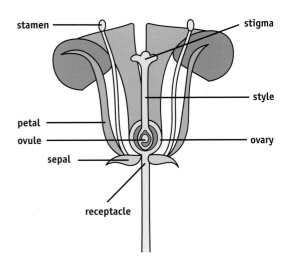

Pollen grains

If you look at pollen grains under a microscope, you will see that they come in many different shapes and sizes. Some are extremely beautiful and look more like exotic jewels or sculptures than plant material. The variation you can see between the pollen grains from different plants is partly a result of the different ways in which the pollen is dispersed. For example the pollen from a grass is dispersed by the wind, so it must be small and light. Pollen that is dispersed by insects is often heavy, sticky and rich in protein to attract insects to it.

The dark orange pollen of this lily flower contains male sperm cells. If the cells reach the egg cell of another flower, fertilisation will occur.

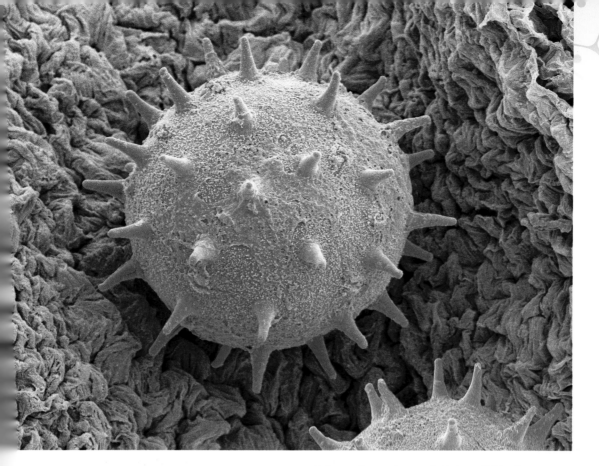

Pollen grains can be amazingly beautiful in shape and colour. This is the pollen grain of a lily.

Inside a pollen grain

Each pollen grain contains two different cells, each with its own job to do. One cell is called the pollen tube cell. When the pollen grain lands on the female part of a flower, it soaks up water and its outer coat softens. The pollen tube cell grows very quickly into a long, thin tube. The tube stretches out from the pollen grain to the female egg cell. The pollen tube cell can do this because it has a store of chemicals needed for rapid growth. The other cell in the pollen grain divides to form two sperm cells. When the pollen tube forms, the sperm cells travel down the pollen tube to the ovule. One fertilises the female egg cell, and the other acts as a food store for the developing embryo.

ANCIENT POLLEN

Pollen can survive for amazingly long periods of time. Scientists who study ancient life forms have taken 'core' samples from the soil. These are long columns of soil that are obtained by drilling into the ground with a long, hollow cylinder. The deeper the sample, the older the material it contains. Some pollen grains from samples in Sweden have been found to be more than 40,000 years old!

Animal cells

Animal cells are enclosed by a a thin layer called the cell membrane. Inside this, organelles such as the nucleus and mitochondria float around inside a fluid called cytoplasm. Different cells are adapted to carry out particular jobs.

Epithelial cells

Epithelial cells line the airways that lead to the lungs. They are adapted to move dust, dirt, germs and mucus away from the lungs and out of the body. They produce a sticky mucus that traps dirt and other particles. Tiny hairs, called cilia, on their surfaces move the dust and mucus along. Mitochondria inside the cells provide the energy to move the cilia.

The 'arms' of this nerve cell branch out to touch and communicate with other nerve cells.

Nerve cells

Nerve cells, or neurons, form vast networks that connect one part of the body to another, usually via the brain. For example, the brain controls muscle movement using motor neurons. Inside the brain, each nerve cell connects with many other neurons. Neurons can carry signals around the body because they have a very long, thin section, called the axon, which transmits the nerve signals.

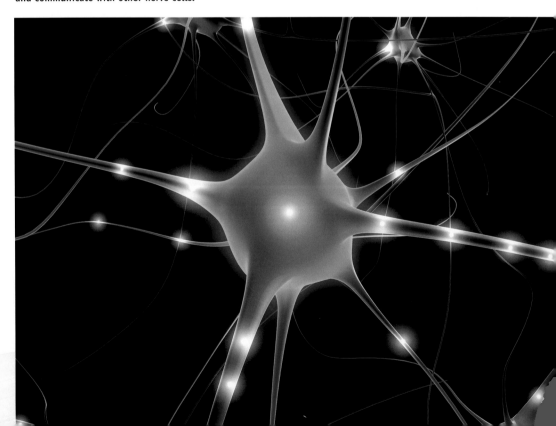

Physical activities use muscles that are made up from millions of individual muscle cells.

At one end of the neuron is the cell body, which contains the nucleus. Small fibres, called dendrites, branch out from the cell body and make connections with the dendrites of other neurons, allowing signals to pass between them.

Rods and cones

Rods and cones are cells in the retina – the light-sensitive layer at the back of the eye. These cells contain light-sensitive chemicals. When light falls on them, the cells respond by sending a signal to the brain via the optic nerve.

MUSCLE CELLS

Muscle cells run along the length of a muscle such as the biceps in your upper arm and the quadriceps in your thigh. They contract and relax, shortening or lengthening the muscle so that it can pull the bones into new positions. Two chemicals – actin and myosin – work together to bring about the contraction and relaxation of muscle cells.

Egg cells and sperm cells

Egg and sperm are special types of animal cells. They form in the sex organs – testes in males and ovaries in females. Sex cells contain only a half set of chromosomes. During fertilisation, the sperm and egg join. Their nuclei combine to make up the full set of chromosomes. The first cell of the new organism has formed.

Egg cells

The egg cells of humans and other mammals are called ova (singular, ovum). Their job is to combine with the sperm cells and develop into new organisms. When the egg and sperm fuse, the complete set of chromosomes is restored (half from each parent). Eggs also have a jelly-like outer layer that allows only one sperm cell to fertilise the egg cell.

Egg cells contain a lot of nutrient-rich cytoplasm. This is full of mitochondria that produce enough energy to support the developing organism until it can obtain nutrients from its mother's body. In humans, the egg supports the developing embryo for about 10 days, until food is passed to the embryo via the umbilical cord.

Here, a single sperm cell is approaching an egg cell, just before fertilisation.

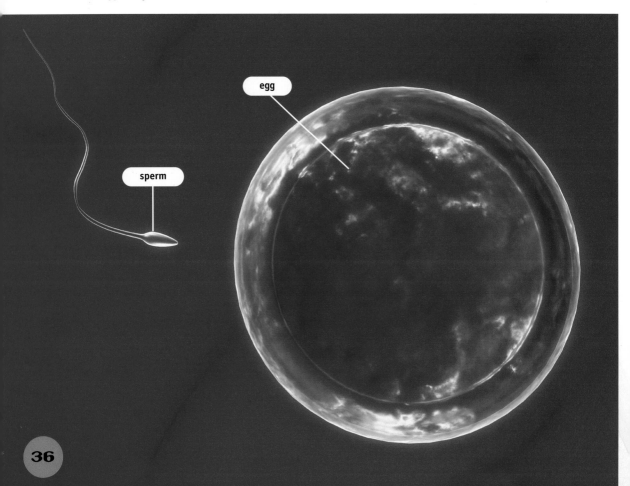

egg

sperm

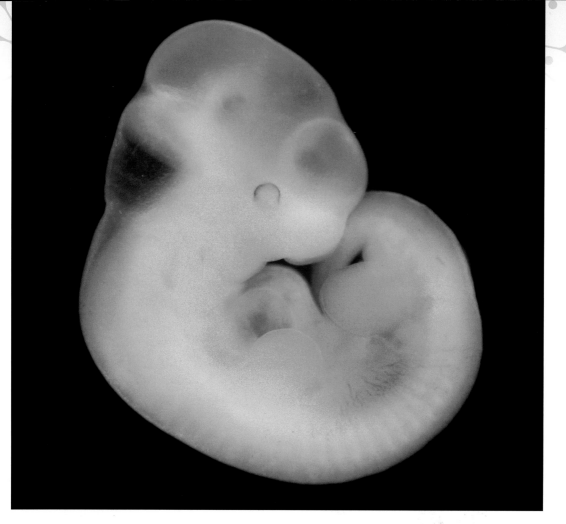

This mouse embryo developed from the fusion of a single sperm cell and egg cell.

Sperm cells

The sperm cells of humans and other mammals are much smaller than egg cells. The sperm cells must swim from the male body to the egg cells. They then join with the egg cells to create new organisms.

Sperm cells have a long tail and streamlined shape, which help them to swim. They also contain many mitochondria to provide energy for swimming. The head section contains chemicals that dissolve part of the egg cell membrane to allow the sperm to enter it.

Millions of sperm cells are released at once to increase the chances of at least one fertilising the egg cell.

FERTILISATION

Millions of sperm may reach the egg cell. Every one will try to penetrate the jelly-like outer layer of the ovum. Chemicals in the head of the sperm attack the cell membrane of the ovum, allowing the head to burrow its way in. As soon as this happens, no other sperm cell can penetrate the ovum. A new life has begun.

Blood cells

In animals, vital functions such as transporting nutrients around the body and fighting infections are carried out by blood cells. These cells are made in bone marrow. Red blood cells are fully formed when they leave bone marrow. White blood cells travel from the bone marrow to other parts of the body, such as the thymus, spleen and lymph nodes, where they mature and develop.

Red blood cells

Each red blood cell is a small, concave disc. Unlike most cells, red blood cells do not have a nucleus. Their main job is to take oxygen from the lungs to parts of the body where it is needed. Red blood cells can do this because they contain haemoglobin. When there is a lot of oxygen in the blood, such as in the blood vessels in the lungs, haemoglobin binds with oxygen to make oxy-haemoglobin. As the oxygen-rich blood circulates around the body, the oxy-haemoglobin breaks down and oxygen is released. The haemoglobin molecule is then ready to bind with more oxygen in the lungs.

White blood cells

There are many different types of white blood cell. Each one has a specific job to do. Together, the white blood cells are an important part of the immune system, which defends the body against disease.

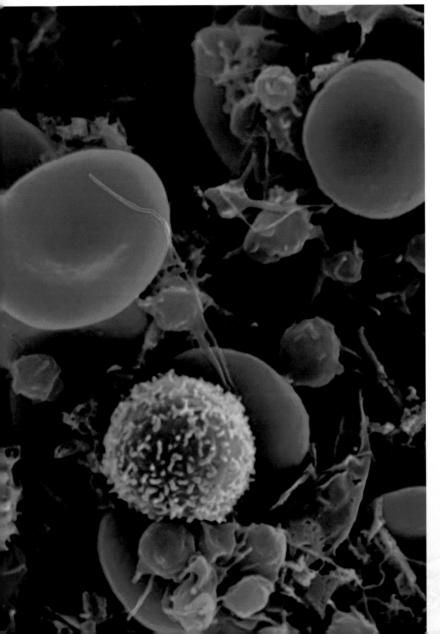

The red circular objects in this photograph are red blood cells. The green-yellow features are white blood cells.

Different white blood cells are adapted in different ways. Some can change their shape to surround and engulf microbes such as bacteria. Chemicals in the white blood cell attack the bacteria, killing them and slowly breaking them down.

Some white blood cells have special chemicals on their surface that 'fit' chemicals on the surface of a microbe. When the white blood cell locks on to a microbe that 'fits' it, it triggers a sequence of reactions in the body that prevent an infection developing.

Some white blood cells can 'remember' a microbe they have encountered before. This means that, if they encounter it again, they can trigger a very swift reaction by the rest of the immune system to prevent an infection developing. It is these cells that make it unusual for us to suffer from illnesses such as chickenpox more than once.

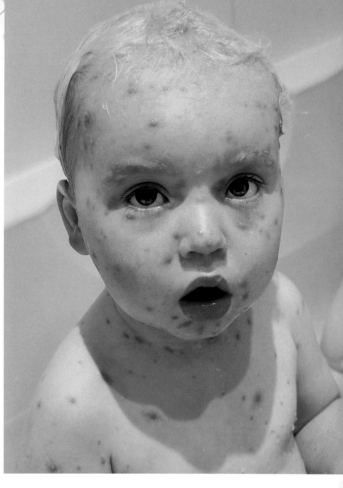

This boy is suffering from a chickenpox infection, but his white blood cells will remember the chickenpox virus and prevent him catching the infection again.

BLOOD CELL TYPES

This table gives information about different types of blood cell in a healthy human. (One micrometre is one-thousandth of a millimetre.)

Cell type	Diameter (micrometres)	Number in 1mm³ of blood	Features	Function
Erythrocyte (red blood cell)	7–8	4–6 million	Red, flattened disc; no nucleus	Transport oxygen
Lymphocyte	6–14	1,700–3,000	Large, round nucleus	Kill microbes and remember them
Neutrophil	10–12	3,500–7,000	Lobed nucleus	Engulf microbes
Monocyte	12–20	100–600	Horseshoe-shaped nucleus	Engulf microbes

Stem cells

Animals and plants contain many different types of cells. Each one develops from a special type of cell called a stem cell.

What are stem cells?

When a cell divides, it produces two daughter cells of the same type. For example, a muscle cell always divides to produce two muscle cells, and a neuron always divides to produce two neurons. Stem cells are different. They have the potential to develop into any type of cell.

Where are stem cells found?

A developing embryo contains many stem cells. Gradually, these stem cells will develop into specialised cells and group together to form tissues. In mammals, the blood of the umbilical cord is also rich in stem cells. Some adult tissues, such as bone marrow, contain stem cells, too, but there are far fewer than in the tissues of embryos. This means that it is harder to use adult tissues as a source of stem cells. Instead, scientists use the blood of the umbilical cord or the tissues of developing embryos as a source of stem cells. This has led to controversy among people who feel that embryos should not be experimented with (see box).

How are stem cells used?

Stem cells have been used to treat illnesses such as leukaemia, which causes white blood cells to malfunction. First, drugs are used to kill the patient's own white blood cells. Next, stem cells from a donor's bone marrow are transplanted into the patient. These grow and divide to form healthy new white blood cells.

Stem cells may be used to treat paralysed patients in the future.

INVESTIGATE:
Right or wrong?

There are both arguments for and against using stem cells for research. Search in newspapers, magazines and on the Internet for some arguments for and against their use. Record your facts in a table like the one below. Then try to decide whether you think it is right or wrong to use these stem cells for research.

Advantages of using embryonic tissue	Advantages of using umbilical cord blood	Disadvantages of using embryonic tissue	Disadvantages of using umbilical cord blood

New organs from stem cells?

One day, scientists hope to use stem cells to grow new organs. Doctors could replace damaged organs with ones grown from stem cells. The patient's body would not reject the transplant, because it is made from the same cells. It would also mean there would be no shortage of organs for transplant patients. This may seem like science fiction, but a piece of windpipe has already been grown in this way and successfully transplanted into a patient.

Claudia Castillo was the first person in the world to have a new organ grown from her own stem cells and transplanted into her body. Her damaged windpipe was replaced with a new one grown from stem cells.

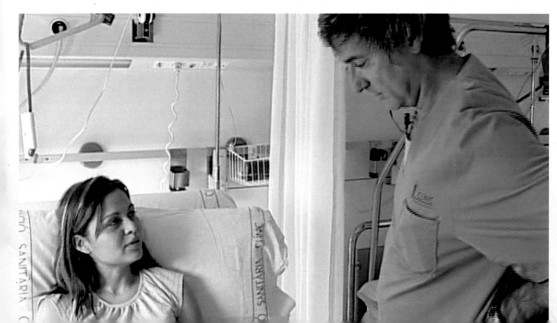

Make a model plant cell

In this book, you have read about specialised plant and animal cells that are adapted in different ways to do specific jobs. Can you remember the basic structure of a plant and animal cell? Try making a plant cell from the inside out.

You will need:

- small round balloon
- large, sealable plastic bag
- lengths of wool
- small, sealable plastic bag
- non-fungicidal wallpaper paste
- water
- green beads
- mixture of beads of different colours
- large box or plastic container

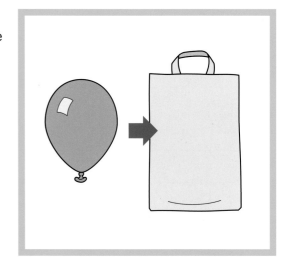

1. Blow up the balloon. Put it into the large bag. It should fill half of the bag. This will be the vacuole.

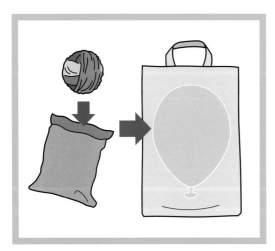

2. Mix the pieces of wool into a rough, tangly ball. This represents the chromosomes. Put it into the small plastic bag and seal it. This represents the nucleus. Put the small bag into the large bag.

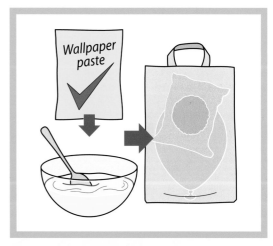

3. Mix the wallpaper paste and water to make a runny liquid. Pour it into the large bag, leaving some space. This represents the cytoplasm.

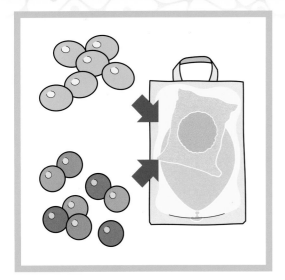

4. Add some green beads. These represent chloroplasts. Add the other beads. These represent other organelles, such as ribosomes and grains of starch.

5. Seal the large bag. This represents the cell membrane.

6. Put the bag and its contents inside the box or plastic container. This represents the cell wall.

Finished – one model plant cell!

MAKING A SPECIALISED CELL

Can you think of ways of adapting your model cell to make it into a model of a specialised cell? Perhaps you could stick lengths of wool or string to the plastic bag 'cell membrane' to represent the cilia of an epithelial cell. What other ideas can you think of?

Glossary

active transport Mechanism by which large molecules can pass through cell membranes.

cell Basic building block of all living organisms.

cell membrane Thin 'skin' enclosing the cytoplasm and organelles in all cells.

cell sap Watery liquid in the vacuole of a plant cell.

cell wall Strong outer layer of a plant cell.

cellular respiration Process by which energy is released in cells.

chemosynthesis Making food from minerals and water.

chlorophyll Green pigment found in the cells of green plants.

chloroplast Organelle in green plant cells that contains chlorophyll. Chloroplasts are the site of photosynthesis.

chromosome Thread-like structure that carries genetic information.

cilia Tiny hair-like structures that line the surface of some cells and help them move.

cytoplasm Fluid inside cells.

diffusion When molecules spread out to even out their distribution in a liquid.

DNA Deoxyribonucleic acid, the chemical from which chromosomes are made.

electron microscope Powerful microscope that uses electrons instead of light to look at objects in fine detail.

endoplasmic reticulum Part of a cell where proteins are made.

enzyme Chemical that speeds up chemical reactions.

flagellae Tail-like structures that help some cells move.

focus Adjust to give a clear image.

gene Sequence of DNA that carries one piece of genetic information.

genome An organism's complete set of genes.

Golgi body Structure inside a cell in which complex molecules are made.

magnification Number of times bigger than actual size.

meiosis Type of cell division that produces four non-identical sex cells.

microscope Instrument used to observe things too small to be seen by the naked eye.

mitochondria Structures inside cells in which energy is stored and released.

mitosis Type of cell division that produces two identical cells.

mutation Change in the genome of an organism.

nucleus Structure inside a cell that contains genetic information.

organ Collection of cells and tissues that work together to do a specific job. The heart, liver and kidneys are examples of organs.

organelle One of a number of different structures found inside cells. Mitochondria and chloroplasts are examples of organelles.

osmosis Process by which water molecules move across a semi-permeable membrane to equalise the concentration of a solution on both sides.

photosynthesis Process by which green plants trap energy from sunlight and use it to make food from carbon dioxide and water.

ribosome Structure inside a cell that makes proteins.

semi-permeable Allowing some, but not all, substances to pass through.

stain Coloured chemical used to make parts of a cell visible under a microscope.

stem cell Cell that can develop into any other type of cell.

tissue Collection of cells that work together to do a specific job. For example, the airways are kept free from dust and mucus by epithelial tissue that is made from epithelial cells.

vacuole Space within a plant cell that contains cell sap.

Further information

WEBSITES TO VISIT

Find out lots of detailed information about cells, together with other biology topics, and test yourself with some fun quizzes at:
www.biology4kids.com/files/cell_main.html

Consult the core curriculum material on cells and life processes at:
www.bbc.co.uk/schools/ks3bitesize/science/biology/cells_intro.shtml

Find out more about cell biology, look at some fun animations and microscope images, and test yourself with some interactive puzzles and quizzes at:
www.cellsalive.com

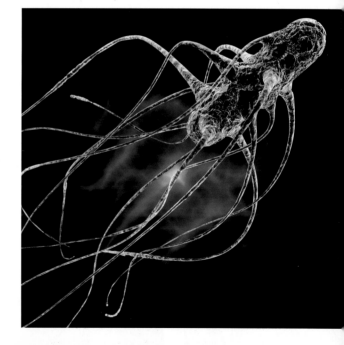

Index